# Join Us

by Lita Davis
illustrated by Noah Jones

**HMH**

Copyright © by Houghton Mifflin Harcourt Publishing Company

All rights reserved. No part of this work may be reproduced or transmitted in any form or by any means, electronic or mechanical, including photocopying or recording, or by any information storage and retrieval system, without the prior written permission of the copyright owner unless such copying is expressly permitted by federal copyright law. Requests for permission to make copies of any part of the work should be submitted through our Permissions website at https://customercare.hmhco.com/contactus/Permissions.html or mailed to Houghton Mifflin Harcourt Publishing Company, Attn: Intellectual Property Licensing, 9400 Southpark Center Loop, Orlando, Florida 32819-8647.

Printed in the U.S.A.

ISBN 978-1-328-77226-8

4 5 6 7 8 9 10  2562  25 24 23 22 21

4500844736          A B C D E F G

If you have received these materials as examination copies free of charge, Houghton Mifflin Harcourt Publishing Company retains title to the materials and they may not be resold. Resale of examination copies is strictly prohibited.

Possession of this publication in print format does not entitle users to convert this publication, or any portion of it, into electronic format.

The girls need more to play.
"Will you join us?" they say.
Yes, the boy will stay.

2　　How many children play?

Two more children come and say,
"May we play with you today?"
"Yes. You can join us. That's okay."

How many children play now?

"We see Ray! Will you join us, Ray?"

"Yes," says Ray.

"I'll stay and play."

4   How many children play?

Three more come to play.
"Can we join you?" they say.
"Yes, you may stay and play."

Now how many children play?

The children stay and play.
No more children come their way.

6   How many children play?

Shep wants to play. "Can I join too?"
Shep seems to say.

How many are playing now?

# Responding

**Problem Solving**

## Stay and Play

**Draw**

Look at page 3. Draw a stick figure for each child on the playground.

**Tell About**

Solve Problems/Make Decisions Look at page 3. Tell how many children you see on the playground. Tell how many children you see coming to join them. Tell how many children there are in all.

**Write**

Look at page 3. Write how many children there are in all.